PROBOSCIS MONKEYS
OF BORNEO

Reproduced from E. Mjoberg's *Forest Life and Adventures in the Malay Archipelago*. George Allen and Unwin, London (1930).

PROBOSCIS MONKEYS
OF BORNEO

Elizabeth L. Bennett and Francis Gombek

NATURAL HISTORY PUBLICATIONS (BORNEO) SDN. BHD.
KOTA KINABALU

&

KOKTAS SABAH BERHAD
RANAU
SABAH, MALAYSIA
1993

NATURAL HISTORY PUBLICATIONS (BORNEO) SDN. BHD.
A928, 9th Floor, Wisma Merdeka,
P.O. Box 13908, 88846 Kota Kinabalu, Sabah, Malaysia

and

KOKTAS SABAH BERHAD
P.O. Box 196,
89308 Ranau, Sabah, Malaysia

PROBOSCIS MONKEYS OF BORNEO

First published 1993

ISBN 983-812-001-4

Colour Separation & Typesetting by
Poly-Sign Scan (M) Sdn. Bhd, Kuala Lumpur.

Printed in Malaysia by
United Selangor Press Sdn. Bhd., Kuala Lumpur.

C O N T E N T S

F O R E W O R D

It gives me great pleasure to introduce this account of proboscis monkeys for two primary reasons. Elizabeth Bennett and Francis Gombek have been and continue to be involved in the documentation and conservation of these special primates of Borneo and this popular account is a product of their interest in providing a comprehensive natural history account so that these animals can be more widely understood and appreciated. Such an effort has not existed until now. Secondly, these primates are a unique group in the wealth of nature abounding in Borneo, fully deserving wider attention in conservation terms, and undoubtedly one of the most interesting and observable animals there can be.

The system of national parks and nature reserves which exists and continues to be expanded in Borneo reflects the endeavour to "capture" as many unique habitats as possible, particularly crucial in the face of receding prime forest areas, as the century draws to a close. More important than just conservation reserves will be the establishment and propagation of an attitude that includes conservation thinking in all human planning. For all this to happen effectively, we must have adequate facts on forest areas and vegetation types, on the distribution and special biologies of plants and animals, and on the potential threats to wildlife and habitats in different situations. One might add that after the specialists have obtained the facts and incorporated them in specialist conservation work, there remains the need to have the basic information made available in wider circles. This account of the proboscis monkeys, surely one of Borneo's most amazing citizens, accomplishes this very ideal.

We do not deny that a study of proboscis monkeys can be interesting and rewarding for its own sake. But putting all the facts together, relating their special biology to their habits and limitations, and then explaining the threats against their survival and possible conservation measures, is another achievement. This account focuses the deserved attention on the proboscis monkey, the population of which has been steadily declining in many parts of Borneo. It is a rare account of one of Borneo's greatest living treasures, which hopefully will be safe because of the efforts of governments and men and women like Francis Gombek and Elizabeth Bennett.

Datuk Lamri Ali
Director,
Sabah Parks

A C K N O W L E D G E M E N T S

Over the "proboscis monkey years", we have received a wide range of support from many people and organizations. The home of the work has been Sarawak's Forest Department, where FG is employed and ELB is based. We are especially grateful to its director, Datuk Leo Chai, for his support, and also to the successive heads of the National Parks and Wildlife Office: Dr Paul Chai, Philip Ngau Jalong and Ngui Siew Kong. ELB's work has been funded by Wildlife Conservation International of the New York Zoological Society and the World Wide Fund for Nature Malaysia; she is now an employee of the former. Additional small grants have come from the Flora and Fauna Preservation Society and the Primate Society of Great Britain. The Royal Malaysian Air Force has been very generous in providing aerial support on many occasions. In terms of people, we would especially like to thank Anthony Sebastian, Rajanathan Rajaratnam and Ramesh Boonratana ("Zimbo") for allowing us to include their data and ideas on proboscis monkeys. Rajanathan also kindly gave us useful comments on the manuscript. Dr Kathy MacKinnon and Dr Junaidi Payne generously provided information on where and how to see proboscis monkeys in Kalimantan and Sabah respectively. We have received particular help over the years from many people, especially Abang Hj. Kassim Abang Morshidi, Laurentius Ambu, Mahedi Andau, Brig, Jen. (U) Datuk Huang Chew Siong, Dr Mikaail Kavanagh, Lady Y.P. McNeice, Dr Clive Marsh, Dr Junaidi Payne, Ken Scriven, Lt. Kol. (U) Sharkawi Hj. Hasbie, Dr Thomas Struhsaker, Dr Mary Pearl, Dr John Robinson, Martha Schwartz and all the staff of the Sarawak Forest Department who have helped with our fieldwork, especially those at Samunsam Wildlife Sanctuary. FG would also like to thank his family for their patience. We would like to thank C.L. Chan for inviting us to write the book, and for being so positive, cheerful and helpful during its production. In finalising the text, we have benefitted from the comments of Dr K.M. Wong of the Sabah Forestry Department.

THE PROBOSCIS MONKEY.

FIRST IMPRESSIONS

When anybody sees a proboscis monkey in the wild for the first time, they are staggered. Not uncommonly, they make a remark such as "I don't believe that animal!". Even seeing them in zoos is not enough to prepare one for seeing these extraordinary animals in their natural habitat, with all the noise and spectacle involved.

Proboscis monkeys have been making spectacular first impressions on people for a long time. Early naturalists could not agree, though, whether the animals were amazingly wonderful or amazingly grotesque. One of the earliest reports of proboscis monkeys in the wild came from British officer Hugh Low. As long ago as 1848, he said that the proboscis monkey "is remarkable for its very long nose; it is a very fine monkey, in size approaching the orang-utan, but much less disgusting in appearance." Another early explorer-naturalist, Beccari, obviously had somewhat mixed feelings about the animals. On the one hand, he said that "the long-nosed ape [is] of singular and ridiculous aspect", but went on "Why amongst all apes...this one should be provided with a long, prominent and fleshy nose, somewhat hooked at its extremity, it is hard to say. According to Darwinian theory, it might possibly be attributed to sexual selection. If such were the case, we might, perhaps, congratulate the monkey on its good taste".

The calls of the proboscis monkey were also a subject of praise. In 1928, the then curator of the Sarawak Museum, Eric Mjoberg, wrote that "the enormous nose is a sounding board that strengthens and deepens the male's vocal powers....The sound is deep and nasal, strongly reminiscent of the bass viol. Possibly, too, there is some aesthetic touch in it, for the females find the sound attractive and crowd round their musically gifted leader".

These early observers were not always so enthusiastic. Hose went on to say of the proboscis monkey: "his appearance is highly ludicrous, he rejoices in a pendulous fleshy nose which droops at the end almost over his mouth. This appendage has no apparent use, and is not even decorative". After he had complimented the monkey's calls, Mjoberg wrote that its nose "puts even the most exaggerated and splendid Bourbon nose into the shade. It is a bright red, fleshy appendage....that projects far above the mouth and partly blocks its entrance. When the owner in question has to satisfy his stomach's insistent

◀ Reproduced from D.G. Elliott *A Review of the Primates, Volume III. Anthropoidea: Miopithecus to Pan*. American Museum of Natural History, New York (1912).

demands, his pushes his 'stop-cock' to one side with his hand, a most comical proceeding". One of Mjoberg's successors as Curator of the Sarawak Museum, the well-known Tom Harrison, referred in 1938 to "the vile porty-looking proboscis monkey", and as late as 1965, a paper was written by the American J.R. Kern entitled "Grotesque honker of the Bornean swamps".

So what is the animal that produces such strong but conflicting reactions really like? And what does it do, that causes such comment? In this short book, we would like to introduce you to proboscis monkeys by telling you about them and their behaviour. We also include discussion about the problems facing them, their conservation and future prospects. We conclude with a short guide to where you can easily see them and how to get there.

Adult male proboscis monkeys are notable for their outsized nose and stomach.

WHAT ARE PROBOSCIS MONKEYS?

Proboscis monkeys belong to the group of animals called Primates. Unlike many other groups of animals, you cannot point to one particular feature of a primate and say right, that is what makes this animal a primate. Instead, primates are characterised by having most or all of a range of features. Most or all primates have hands and feet well adapted for grasping objects, with separated and very mobile fingers and toes, and nails rather than claws. They also have their eyes at the front of their heads, not round the sides, so have good stereoscopic sight like humans. And they generally have relatively larger brains than other animals.

The order Primates includes humans, as well as apes (chimpanzees and gorillas of Africa, and orang-utans and gibbons of Asia), all the monkeys, and a group of smaller, generally nocturnal animals known as prosimians. Monkeys of the Old World (Africa and Eurasia) and New World (America) are very different and not especially closely related. Only New World monkeys have prehensile or clinging tails, for example.

In the Old World, monkeys fall clearly into two groups known as cercopithecines and colobines. Proboscis monkeys are colobines. They say a way to a man's heart is through his stomach, so with a colobine, it is a hard task indeed since colobines have the most enormous stomachs. It is this which distinguishes colobines from cercopithecine monkeys such as macaques and baboons. A colobine's stomach is divided into several sections, similar to that of a cow. Like a cow's, a colobine's stomach is full of a soup containing huge numbers of bacteria which ferment the animal's food. This allows it to digest leaves to obtain energy; animals with normal, simple stomachs, like humans, cannot do this.

The ability to obtain energy by eating leaves means that, in terms of numbers, colobines are by far the most successful primates in tropical rain forests of Africa and Asia; in any one forest, about two-thirds of the primates are colobines. In African forests, these are colobus monkeys, and in Asia, they are the langurs or leaf monkeys of South and South-east Asia, such as the banded, silvered and red langurs, as well as the beautiful and exotic looking douc and golden monkeys of Indochina and China. The latter two species, together with the simakobu of the Mentawai Islands near Sumatra, all have rather unusual noses and, with the proboscis monkey, are known as "odd-nosed colobines". None of the others has a nose in the same outsize class as the proboscis monkey, but they are unusual and upturned, and these are the proboscis monkeys' closest relatives.

Golden Monkey

Proboscis Monkey

Simakobu

Douc

"ODD-NOSED COLOBINES"

The colobines' specialised stomachs and adaptations for living in forests mean they do not do so well in more open, or man-made environments. Here, the more versatile cercopithecines come into their own, being able to eat a wider variety of items. So it is the cercopithecines with which we are generally more familiar. That is not because they are more common overall; it is that they tend to thrive in areas where Man is. In the forests, colobines predominate.

Proboscis monkeys are particularly extreme, even for colobines. Their stomachs are relatively twice as large as those of any of their relatives. This means that, to the uninitiated, proboscis monkeys appear permanently pregnant - even the males! Their enormous, bloated stomachs and huge, pendulous noses have given rise to one of their less flattering names, *orang belanda* or 'Dutchman', since the animals reminded people in Borneo of Europeans. *Orang belanda* is still the name most commonly used for the animals in Sarawak, especially in the west. In other parts of Borneo, a wide range of names is used for the animals, including *rasong, raseng, pika, bekantan, bentangan* and *bangketan*.

It is the adult male proboscis monkey which is so extreme and striking. Weighing an average of 20 kg, he is very large for a tree-dwelling animal, and indeed, he is one of the largest monkeys in the world. He has a flesh coloured face, and his huge, pendulous and greatly expanded fleshy nose does indeed overhang his mouth, forcing him to push it up out of the way when he eats. Proboscis monkeys are mainly reddish-brown, with greyish limbs. The male has a darker cap over the top of his head, a yellow collar and a thick, dark brown mane of fur on his back which resembles an old bomber jacket acquired in his youth and into which he is now trying to cram his middle-aged spread. At the back below his short jacket, he has a strikingly white rump patch of much shorter fur, leading into a thick white tail.

Although she shares most of the same basic features as the male, a female proboscis monkey is altogether a more modest looking animal. She is only about half his weight, around 10 kg. Like the male, she is also mainly red-brown with a flesh-coloured face, but she does not have his striking contrasts of colour. Moreover, her nose is not huge and pendulous. In younger females, it is fairly short and snubby, and even noses of older females which droop somewhat never reach the enormous dimensions of the male's and never reach as low as the mouth. This means that it is easy to pick out a male from any angle. Even if the only thing visible through the leaves is a row of tails, the male is the one whose tail is thickest and whitest.

New-born proboscis monkeys look quite different from the adults, but are also striking. They are covered with sparse, blackish fur and have blue faces with snubby, upturned noses. Their fur turns brown quite quickly, and by the time the monkey is about four months old, it is roughly the same colour as an adult's. The face slowly turns greyish, but does not lose its bluish tinge totally until it is about a year old.

Not only are proboscis monkeys unusual to look at; they also make the most bizarre range of noises. Roars, grunts, squeals and a huge range of nasal honks mean that, from a distance, a group of proboscis monkeys through the trees can sound somewhat akin to a group of grumbling old men. And the occasional roars sound like a far larger and more dangerous animal than a monkey! Most commonly, different proboscis monkeys in a group start to honk and squeal, then the male gives a gentle and deep "ho-hoooong", which seems to calm them down and peace is restored.

Typical habitat of the proboscis monkey: mangrove lining a river in the coastal plain.

WHERE ARE PROBOSCIS MONKEYS FOUND?

The only place in the world where proboscis monkeys occur is the island of Borneo in South-east Asia. And they are not even found throughout all of Borneo. They are forest-dwellers and are limited mainly to coastal swamp forests and to forests next to large rivers, again, generally not far inland. Even more limiting, they are not found in all areas of coastal swamp forest. These swamps mainly comprise mangrove and peat swamp forests. Both of these habitats have different zones, where different types of trees occur somewhat patchily, depending on distance from the coast or centre of the peat swamp. Proboscis monkeys are found in most of these forest zones, but not all. For example, it is uncommon to find them in large patches of nipa mangrove; extensive stands of the brackish-water nipa palm form an inhospitable area with little food for a monkey. On the other hand, in some forests next to rivers, they reach very high densities. So it is misleading to assume that any area of mangrove or peat swamp forest is full of proboscis monkeys. Only by going there to look can you be sure.

Although proboscis monkeys are mainly restricted to coastal areas, small numbers are occasionally found much further inland next to major rivers such as the Segama and the Kinabatangan in Sabah and Barito in Kalimantan. There are even scanty reports of animals passing briefly through hill forests far in the interior. This is extremely rare, and such animals are probably on their own and nomadic.

Nobody knows why proboscis monkeys have such a very limited distribution, and why they are not found in the vast tracts of rain forest throughout inland Borneo. It is possibly related to the lack of minerals or other nutrients. Inland forests of Borneo grow on notoriously poor soils. Proboscis monkeys are large animals so they need an ample supply of digestible food. This is most likely to be found where forests grow on nutrient-rich alluvial soils such as mangroves or alongside rivers.

BORNEAN RESERVES WHERE PROBOSCIS MONKEYS ARE FOUND

Being restricted mainly to coastal areas, proboscis monkeys are not found in many of the large, inland reserves of Borneo. Borneo has coastal reserves, but most of them are small, and the monkeys often use areas inside and outside the reserve. Reserves which are too small to protect a viable population of proboscis monkeys if all the forest surrounding them were cleared are marked with a *.

Throughout the entire range of the proboscis monkey, then, the only reserves known definitely to be large enough to sustain a population if isolated are Tanjung Puting and possibly Gunung Palung National Parks in Kalimantan.

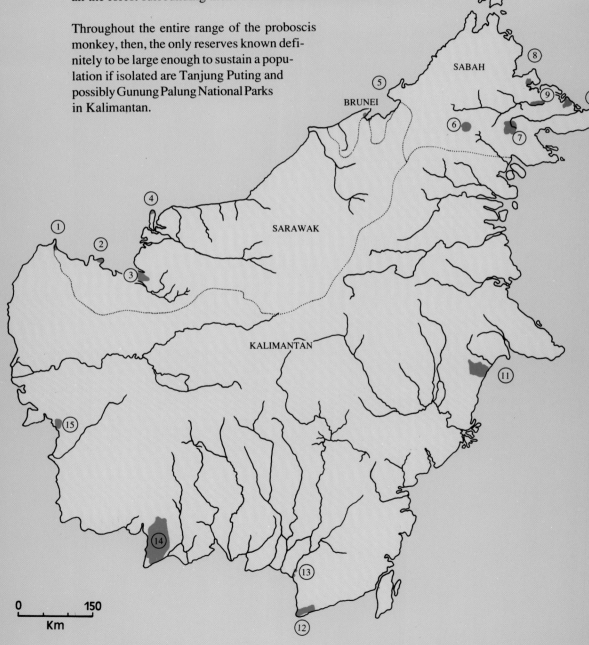

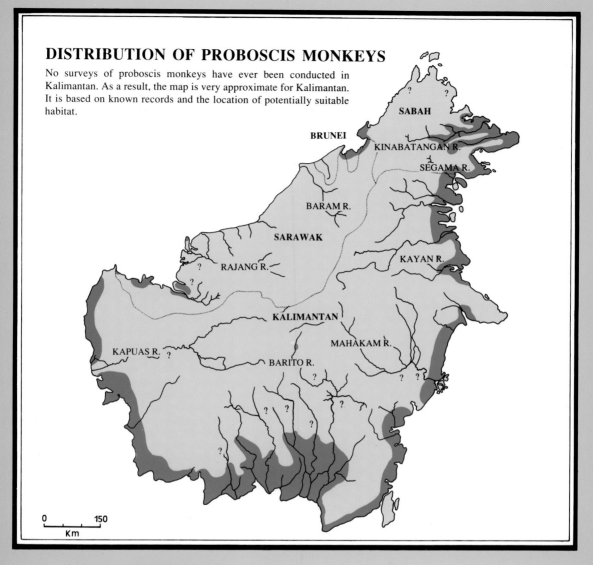

DISTRIBUTION OF PROBOSCIS MONKEYS

No survey of proboscis monkeys have ever been conducted in
Kalimantan. As a result, the map is very approximate for Kalimantan.
It is based on known records and the location of potentially suitable
habitat.

SABAH

BRUNEI

KINABATANGAN R.

SEGAMA R.

BARAM R.

SARAWAK

KAYAN R.

RAJANG R.

KALIMANTAN

MAHAKAM R.

KAPUAS R.

BARITO R.

0   150
Km

1. Samunsam Wildlife Sanctuary *
2. Bako National Park *
3. Proposed Maludam Wildlife Santuary *
4. Proposed Bruit-Patok National Park *
5. Pulau Siarau & Pulau Beramban Primary Conservation Areas *
6. Maliau Basin Conservation Area *
7. Danum Valley Conservation Area *
8. Sepilok Forest Reserve *
9. Proposed Lower Kinabatangan National Park
10. Kulamba Wildlife Reserve *
11. Kutai National Park *
12. Pleihari Tana Laut Wildlife Reserve *
13. Pulau Kembang & Pulau Kaget Wildlife Reserves *
14. Tanjung Puting National Park
15. Gunung Palung National Park

Females and young of a proboscis monkey harem group.

THE SOCIAL LIFE OF PROBOSCIS MONKEYS

When you first look at proboscis monkeys in the wild, it seems that their social life is chaotic. Sometimes there are just one or two animals, sometimes about ten, and sometimes huge, noisy hordes of 80 or more. Early observers could not agree on what was happening, whether the animals lived in small groups with only one male, or if they were in much larger ones, with many males and females. The impression was that there was no set social system, merely random associations of monkeys. That would be highly unusual. No animals live in a state of social chaos, especially not higher primates which have strict patterns and rules governing their social lives.

In fact, proboscis monkeys do have a set social system, but it is more flexible, and possibly more subtle, than that of many other primates. They live in harems which are groups containing one male, one to eight females and their offspring, the average harem size being about nine animals.

Female proboscis monkeys with a blue-faced infant, probably about nine months old.

With both male and female proboscis monkeys moving between harems at different times, the closest social unit is a female and her offspring.

▲ Infants remain close to their mothers for up to two years.

◄ A successful male proboscis monkey has to attract females for his harem and defend them against rival males. Physical prowess is essential.

Living in harems is not unusual - most primates and many other mammals have social systems based in some way round a harem. What is unusual about proboscis monkeys is that the harems do not stay apart and avoid others. They frequently come together, especially by riversides in the evening. This explains the large hordes of animals sometimes spotted.

The other unusual thing about proboscis monkeys is the frequency with which animals move between different social groups. Usually, animals stay in the social group they know. If one sex moves between groups to avoid inbreeding, the other stays behind with its relatives. However, both male and female proboscis monkeys sometimes move between social groups. Females switch between harems at any time from before adolescence onwards, and they might change groups several times in their lives. Males are kicked out of the group in which

▲ Young proboscis monkeys spend much of their time playing, either on their own or with others.

▶ Adolescent female proboscis monkeys sometimes move to a new harem before breeding.

Proboscis monkey harems contain an average of nine animals, but can number up to twenty.

Female proboscis monkeys frequently groom their offspring. This removes dirt and parasites, and provides social bonding and support.

▲ Harem male proboscis monkeys frequently climb to high vantage points...

▶ ... where they can impress the females, deter potential rival males, and be on the lookout for both.

Juvenile male proboscis monkeys either live on their own or join with others to form an all-male group.

they were born when they are only just old enough to fend for themselves, maybe when they are only one to two years old. They team up with other males to form all-male groups. These typically comprise one or two large males and about ten small ones. This is misleading for the unwary observer because they look very like harems, with one large animal and many smaller ones. The illusion of a harem is enhanced by young males clinging to older ones to sleep at night, in much the same way as they would have clung to their mothers. All-male groups also congregate with harems along the riverbanks, and often follow harems as they travel through the forest during the day.

For a male proboscis monkey, coming close to other harems and all-male groups means that other males are sometimes close to his females. This is a major problem. Either the females might be enticed away, or another male might sneak in for a quick mating while his back is turned. So, if another group is nearby, a harem male often gives a highly spectacular display of strength. He is trying

◀ Young females have to assess the merits of the different harem males before deciding with whom to mate.

▶▶ An unusual sight: two adult males travelling together. They are part of an all-male group.

21

to show his, and possibly other, females how big, strong and handsome he is. At the same time, he is telling other males that he is a formidable opponent so stay away from his females!

The first stage of a display involves a male staring hard at his opponent, leaning forward on all fours with his chin thrust forwards or his mouth wide open. If that does not have the desired effect, the male then suddenly and unpredictably leaps through the trees, often with a loud roar, and frequently landing on dead branches which break with a sharp crack, adding to the general uproar. It is rare for confrontations between males to escalate to actual physical contact. Chases between males are not infrequent though, and sometimes one male will chase another so suddenly that the latter leaps off his branch and lands with a loud splash in the river below. If the males do actually make contact, they have slapping contests. Judging from occasional scars, it seems that males sometimes get injured in fights, but it is probably quite rare. Showing their strength using displays means that males can assess their chances of winning a battle, so do not attempt one if their chances of success are low or if sustaining injury seems likely.

Once harmony with other groups has been restored, proboscis monkeys can resume their social life inside their own group. Compared to some monkeys such as macaques, social life within the group is pretty quiet. Most interactions between animals are infants and juveniles playing with each other, having chasing, grappling and swinging games through the branches. Adult females spend some time grooming their infants but, unlike many other primates, grooming between adults is rare. The only physical interaction often seen between the adult male and his females is mating, and even that is inclined to be seasonal, with a peak around the middle of the year. This corresponds with a peak in births around the turn of the year. Although much smaller than the male, a female proboscis monkey is not shy of him and sometimes initiates the mating. She does this by presenting her back to the male, leaning forward on all fours, and turning her head over her shoulder to face him, waggling her head from side to side and pouting her lips.

Even when he has secured his females from other males and one is presenting herself before him, a harem male faces problems. When mating starts, the young animals in the group become extremely upset and do everything they can to interfere. They frequently pull hard on the male's upper leg, screaming all the while, but a more successful tactic is to lean over the amorous couple from the front and try to tweak the male's nose. Even if this does not stop mating immediately, it certainly curtails a male's ardour. He sometimes even has to stop what he is doing to chase away the youngsters before returning to his female. The ultimate frustration is when he finds that, in the meantime, the female has lost interest and wandered away.

When travelling, harems are usually led by the females.

BIG NOSES

The most common question that springs to most people's minds about proboscis monkeys is why the male has such a huge nose. Answers to this over the years have been many and varied. Proboscis monkeys are proficient swimmers, so it was suggested that the nose acts as a snorkel to help the animals breathe when swimming. This fails to explain why females, in that case, do not drown. Another theory was that the nose is mainly a function of body size: the larger the male, the larger the nose. This does not explain why the nose grows relatively faster than the rest of the body, and is also negated by the fact that when a young male proboscis monkey grows, his body is full sized before his nose expands fully.

Another common misconception is that the nose "balloons" up when the monkey calls, amplifying resonant alarm honks. The nose does move forward slightly when a male calls, but that is purely a function of its being in front of the mouth and the male pushing his chin up when calling. The nose does not increase in size.

The reason why both sexes have larger noses than other primates might have something to do with regulating body temperature. Coastal swamp forests are hot, and proboscis monkeys are larger than any other monkeys living there so have a greater problem of heat loss. They also have large, fermenting chamber stomachs inside them which must generate a lot of heat. So they could well suffer from overheating. A big nose provides a large surface area from which to lose heat, in much the same way as the ears of an elephant do. This could explain why proboscis monkeys have big noses in the first place, but not why the male's carries on growing after he has reached his full body size.

The most likely explanation for the male's nose was hit upon by Beccari as long ago as 1904, when he attributed it to Darwin's concept of sexual selection. Females might quite simply prefer to mate with males with big noses. If so, then males with larger noses will have more females, and therefore more offspring, than males with small noses. In which case, genes for big noses would spread throughout the population. This is much the same concept as why peacocks have large tails - females choose to mate with males which have big, spectacular tails.

► Many bizarre suggestions have been made to explain why male proboscis monkeys have such huge noses.

To us, perhaps, a peacock's tail is a rather more attractive adornment than a proboscis monkey's nose, but there is no accounting for taste, and female proboscis monkeys apparently find long, pendulous noses to be the ultimate in good looks.

The male proboscis monkey's nose is so large, he sometimes has to push it out of the way when he eats.

LARGE STOMACHS, THEIR USES AND HAZARDS

Proboscis monkeys have greatly enlarged, cow-like stomachs. These contain a vast array of bacteria which ferment the animal's food. Unlike cows, proboscis monkeys do not chew the cud but their stomachs, like those of cows, have two big advantages. First, the bacteria can break down cellulose, which is the main structural part of leaves. So proboscis monkeys can obtain energy from leaves, whereas cercopithecine monkeys, apes and humans cannot. This goes a long way to explaining why colobine monkeys are so successful in forests. The second advantage of the system is that the bacteria deactivate at least some poisons in the food, allowing the monkeys to eat nasty foods which would kill any normal animal.

Like everything, the system does have its disadvantages too. If the colobine monkey eats highly digestible foods such as sweet, sugary fruits, the bacteria ferment them so rapidly that gas and acid build up suddenly in the stomach. This highly uncomfortable sounding condition known as bloat can quickly kill the animal. So proboscis monkeys have to forego tasty, sweet fruits such as figs and rambutans, and stick to bitter, tough ones, most of which are pretty unpalatable to us humans with our simple stomachs.

The mangrove, peat swamp and riverine forests of Borneo are always lush and green, with a superabundance of leaves in all directions all year round. This is paradise, you would think, for a monkey adapted to eating leaves. Staggeringly, there is so little food in many areas that the monkeys have trouble finding enough to survive.

The reason is that the trees have taken steps to protect themselves against the depredations of hungry animals. Some such as rattans and nibong palms make themselves physically unpleasant, producing huge barrages of long spines and thorns. A more widespread tactic of the trees is to contain chemicals which are harmful to animals. Proboscis monkeys and other colobines generally cope with this better than other animals, since their stomach bacteria can deactivate certain poisons. On the other hand, if the plants contain lots of fibre and tannin which impede digestion, the monkeys quite simply cannot digest them. Moreover, some plants contain chemicals that act as antibiotics. Proboscis monkeys cannot eat these either because they would kill the bacteria in their stomachs; without the bacteria, the monkey could not eat many of its other foods, so it has to nurture the bacteria and avoid eating anything which harms them.

Most of the mature leaves in these coastal swamp forests are extremely tough and fibrous, which means that they are not available as a food source for the monkeys. Instead, proboscis monkeys have to resort to young leaves, which are less fibrous but are far scarcer and more scattered in the forest. During the course of a year, about half of the animal's diet is made up of young leaves with just a few mature leaves.

Proboscis monkeys cannot top up their diet with sweet, succulent fruits either, because of the problem of bloat. So the other half of their diet is made up of non-sweet fruits and seeds, such as those of the nutmeg, legume and palm families. These too are often hard to find, especially during certain months of the year when few trees are producing fruits. So all in all, proboscis monkeys are faced with "leaves, leaves everywhere but not a lot to eat".

This diet is typical of colobine monkeys, mixing young leaves, non-sweet fruits and seeds. It is a very different diet to that of a primate with a simple stomach. Orang-utans and macaques cannot get energy from young leaves, but they can

▲ The expanded stomachs allow proboscis monkeys to eat tough and poisonous foods, but force them to avoid sweet succulent ones.

◀ Proboscis monkeys have greatly expanded, chambered stomachs in which their food is fermented by a myriad of bacteria.

31

Their large stomachs mean that proboscis monkeys look permanently pregnant – even if they are obviously male.

eat succulent fruits without fear of bloat. So their diet is made up of sweet fruits such as figs and a wide variety of other wild fruits, supplemented by a few tender young leaves and a variety of other items such as termites.

The different digestive systems of the primates means, then, that they cannot eat many of the same foods as each other. Gibbons would poison themselves if they ate many of the legume seeds gorged on by proboscis monkeys, and proboscis monkeys would die from bloat if they had a fig feast. It also explains why, more often than not, if a group of noisy, small macaques crashes through a group of more sedentary proboscis monkeys, the two species largely ignore each other. They are looking for different foods, are not competing, and therefore have no need to interact.

Functions performed by having a complex stomach (proboscis monkeys and other colobines) compared to a simple one (humans, apes, monkeys). + means the stomach type can do that; − means the stomach type cannot do that.

	Complex stomach	Simple stomach
Obtain energy from leaves	+	−
Deactivate poisons	+	−
Make some vitamins	+	−
Re-cycle nitrogen	+	−
Digest sweet, energy-rich fruits	−	+
Digest rich easily-accessible protein, e.g., insects	−	+

TRAVEL – PROBOSCIS MONKEYS ON THE MOVE

Daily travel patterns of proboscis monkeys are determined by two main factors - the location of good food sources, and rivers. The animals sleep in trees adjacent to rivers every single night. After dawn, they usually go into the forest away from the river in search of food, although occasionally they amble through the trees near the river for the whole day. Their food is often scarce and scattered, so these large animals have to travel long distances to get enough of it. In fact, their food is so scarce that proboscis monkeys travel further each day than most other species of forest monkey. On some days they go up to 2 km, whereas other forest colobines average 800 m or less. During the course of a year, all other rain forest colobines use an area less than 1 sq km in size. Proboscis monkeys travel much further than this to find enough food, and each group uses about 9 sq km during a year. This area, known as the home range of the group, is often long and thin since the animals rarely go more than 600 m from a river, so their home range is a strip of forest along both sides of the river. In areas with several rivers, the home range is a series of interconnected strips along the different waterways.

Rivers have a major influence on travel patterns of proboscis monkeys.

34

Proboscis monkeys travel almost entirely in the trees...

... only occasionally dropping close to the ground to travel through nipa plams or, even more infrequently, across open ground.

The home range of a group of proboscis monkeys often spans different forest types. The animals might, for example, feed on seeds in riverine forest for part of the year, and migrate downriver into mangrove forest to feed on young leaves there when food in the riverine forest is scarce. This is exactly what happens in Samunsam Wildlife Sanctuary in Sarawak.

Different groups of proboscis monkeys use exactly the same areas as each other. This is quite unusual, since animals often exclude others from their home ranges. But each group of proboscis monkeys has to travel so far to find food, it cannot possibly defend all that area against others and it does not attempt to do so.

An animal that travels widely in areas riddled with networks of rivers and creeks has to be able to swim, and proboscis monkeys are very good at it. They have partly webbed back feet, which help them to swim, and also to walk on mangrove mud without sinking in. They swim using a sort of doggy paddle, moving slowly but with the absolute minimum of noise and splashing. At Samunsam in Sarawak, the animals slide extremely quietly into the water and swim across in single file with no noise or splashing at all. It was assumed that

Imagine an infant's surprise when it finds itself clinging to its mother first in mid-air, then landing in the water with a colossal splash.

Proboscis monkeys are proficient swimmers, moving quietly through the water using a form of "doggy paddle".

this was to avoid attracting estuarine crocodiles, which are the main predators of proboscis monkeys; crocodiles are attracted to prey by splashing. It was a surprise, then, to find that proboscis monkeys in the Kinabatangan area in Sabah cross rivers using an entirely different technique. They climb high into trees by the river's edge, then from about 10 m up, hurl themselves out and down to land in the river with a colossal splash. This area also has crocodiles, but on the other hand, the river is far narrower, so a single wild leap takes the animal two-thirds of the way across, with only a couple of metres to swim. Maybe a much shorter time in the crocodile's river compensates for making such a commotion.

Apart from crocodiles, the other problem of swimming is faced by infant proboscis monkeys. They travel by clinging to their mother's front. Imagine their surprise when they suddenly find themselves hurtling through the air and then plunged into water and unable to breathe! After a few seconds underwater, the infant clambers further up its mother's front, so a little face appears over her shoulder as the infant comes up for air.

Scarcity of food explains why proboscis monkeys have to travel so far and need to cross rivers. But it does not explain why they sleep right next to a river every single night. It is not because that is where all the food is; often the best feeding

Proboscis monkeys invariably travel to riversides every evening, and spend the night sleeping in trees adjacent to or overhanging the water.

trees are a short distance away from the river. So what is it about the river that makes it such a great place to spend the night?

Early observers assumed that proboscis monkeys sleep by the river to avoid predators; that the river on one side, and water below, offered protection from the clouded leopard, Borneo's largest cat. This is not likely, though, because the estuarine crocodile probably eats proboscis monkeys more often than do clouded leopards, and it is not wise to avoid a minor threat by sitting above a major one. Also, if it is a good way to avoid predators, all of the other primates should sit by the river at night, which they do not. Another possibility was that it was cooler by the river at night. With their large stomachs brewing away, and large body size, proboscis monkeys might try to keep cool at night. This might be easier by a river with its circulating air currents. This was tested by hanging thermometers by the river and inland, but the temperatures were no different. Another theory down the drain!

In which case, why do proboscis monkeys sleep by the river at night? It is probably related to their unusual social lives. Females move between harems. In order to decide whether or not they should move, and where they should go, they need to size up the different males, to see which is the strongest, fittest, and likely to father the best children. Riversides are obvious places for the females to have a good look at the males because they are open with good visibility. Not only can the females size up the males, but the good views also mean that riversides are ideal places for males to display and show off their prowess. They also allow males to keep a better eye on their rivals.

So we can now revise our picture and say that a proboscis monkey's travel is influenced by two factors: food and the demands of its somewhat strenuous social life.

▶ Travel between trees often involves spectacular leaps.

NATURAL PREDATORS OF PROBOSCIS MONKEYS

The proboscis monkey is lucky in that no big cats occur in Borneo. There are no tigers or leopards here, and the largest cat is the clouded leopard. This is not likely to take proboscis monkeys because it hunts mainly on the ground and proboscis monkeys live mainly in trees. The only animals which might take proboscis monkeys in trees are eagles, kites and reticulated pythons. Nobody has ever reported seeing proboscis monkeys being eaten by any of these. There is no eagle or kite in Borneo large enough to take adult proboscis monkeys, although they might presumably take the occasional youngster. Reticulated pythons are known to enter cages and eat captive primates, and they grow up to almost 10 m in length, so are perfectly capable of taking proboscis monkeys.

The animal by far the most likely to eat proboscis monkeys is the estuarine crocodile. This is a voracious killer of anything that moves in rivers, and within Borneo, estuarine crocodiles and proboscis monkeys generally live in the same sites, mangroves and riverine areas close to the coast. Again, nobody has ever seen them eating proboscis monkeys, but they kill several people every year and are obviously capable of taking large monkeys. The monkeys are most vulnerable when walking on mangrove mud or swimming across the river, where they have little defence against such a formidable predator. These days, the number of crocodiles has declined considerably, mainly due to hunting by humans.

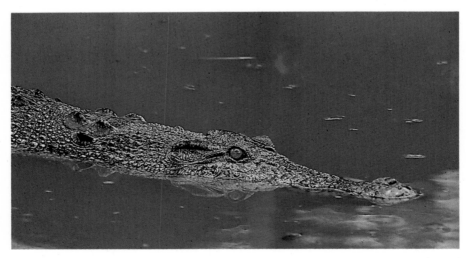

Estuarine crocodile: voracious killer in the rivers, and likely to be the main natural predator of proboscis monkeys.

A DAY IN THE LIFE OF A PROBOSCIS MONKEY

Proboscis monkeys are active in their swamp and riverine forests during the day, and sleep in trees adjacent to a river at night. In general, they are not such early risers as the other primates there which are up and away almost as soon as dawn has broken. Proboscis monkeys usually wake within half an hour of dawn, but sometimes do not leave their sleeping trees for some two hours more. Most commonly, they move away by an hour after dawn, but it is inclined to be earlier if there are many other groups of proboscis monkeys nearby, and later if it is raining.

On waking in the mornings, the first main event of the day is breakfast. Proboscis monkeys do not usually sleep in a good food tree, so they have to move at least a short distance to eat. They sometimes remain close to the river, swim across it or, as often as not, move a short distance away from it into the forest. Then follows a general bout of feeding. Frequently, most of the group sits in one tree. Even so, it is rare to see more than two or three animals at a time due to the denseness of the riverine and mangrove trees; an infant squealing and

The first main event of a proboscis monkey's day is breakfast.

The day is made up of bouts of feeding interspersed by long rest periods.

The main foods of the proboscis monkey are young leaves, shoots and seeds.

In the evenings as groups approach the river, most of the social interactions occur, such as adult males threatening their juniors.

crashing through the branches in play, the occasional fruits dropping from above and the sight of a few tails from time to time are the only signs to any human below that the animals are there.

After the first main feed of the day, the proboscis monkeys usually (but not invariably) move further from the river. Surprisingly, perhaps, the male is not the leader of the group. The first animal to move off is usually an adult female who then leads the group, while the male is inclined to tag along near to the back.

After food and travel comes a rest. All colobine monkeys have long rests during the day, while they digest the leaves or other tough foods they have eaten. Proboscis monkeys, with their unusually large stomachs, carry this to extremes. Once at Samunsam, they had breakfast, moved inland into a tree to rest at 7.50 a.m., then stayed in the same tree for eight hours before moving off for a large meal before dusk. Somewhat tedious for the scientist under the tree! A post-breakfast rest of one to two hours is more common, though, and the animals spend the remainder of the morning and most of the afternoon alternating between feeding, resting and travelling. About two hours before dusk, the

Male proboscis monkey feeding in the late afternoon sunshine.

monkeys start to move back towards the river and have one more feeding session before darkness falls.

As the proboscis monkeys approach the river, the different groups often come together, and it is now that the animals are most active. From time to time, an adult male displays, so the gentle evening sounds of birds and insects are shattered by roaring and branch crashing. Meanwhile, the females carry on feeding or grooming their infants, and the juveniles play, chasing and wrestling with each other in the beautiful, clear evening light. With several groups assembling by the river, such activity is enormously spectacular under the bright orange and pink evening equatorial sky.

Not all evening gatherings comprise such noisy hordes. Sometimes a group potters down to the river on its own and settles with the minimum of noise and fuss.

By dusk, the monkeys have moved up into a suitable sleeping tree and are settling down for the night. Sleeping trees are generally tall and fairly open so visibility is good. This presumably allows the animals to keep an eye out for predators, as well as for males from other groups sneaking in during the night. It possibly also allows them a good overview of where the best place is for breakfast. By the time the nightjars are swooping around and the bats hunting for insects low over the river, the proboscis monkeys are asleep.

Not all days are exactly the same. On some days, the monkeys spend very little time resting and seem to be on the move all day, whereas on others they sleep for most of the time between the post-dawn and pre-dusk feeding bouts. Sometimes they meet other groups, sometimes they do not. All of this is influenced by the amount and distribution of food available in the forest at the time, since that ultimately controls the daily lives of all animals.

PROBOSCIS MONKEYS IN CAPTIVITY

Proboscis monkeys do not generally do well in captivity. They are extremely sensitive, appear to get depressed after a few days, stop eating and die. This depression could be because of poor captive diets upsetting their delicate digestive system, or because of stress. In any case, they rarely survive long once taken from the wild.

There are a very few zoos which have kept them for long periods, such as the Bronx Zoo in New York where they breed extremely successfully. Here, they are kept in a large enclosure, built to resemble a mangrove with artificial mangrove trees adjacent to water, and a carefully controlled diet which even includes mangrove leaves from time to time. Only in such good conditions do proboscis monkeys thrive and breed in captivity.

Unfortunately, proboscis monkeys are such spectacular animals that many zoos want to keep them, whether or not they have the correct facilities and expertise to do so. Since there are so few proboscis monkeys in captivity, zoos that want them have to take them from the wild. In almost all cases, this would be disastrous for wild populations. Proboscis monkeys live in harems. Taking an adult male away disrupts breeding of an entire social group. In Sarawak at least, the largest single population of proboscis monkeys has only about ten harems, therefore ten breeding males. Removing even one of them would stop breeding and disrupt the social life of 10% of the entire population. Moreover, with the very high mortality rate in captivity, at least two animals would probably need to be caught for every one which ends up in a zoo. So the effect on already tiny wild populations would be devastating.

Fortunately, the wildlife authorities in all regions of Borneo are fully aware of the problems and have refused permission to capture proboscis monkeys from the wild in almost all cases. In recent years, virtually the only exceptions have been doomed animals in isolated areas where all the trees were to be cut down for agriculture or other schemes.

Under those circumstances, and with extremely careful and expensive management, proboscis monkeys in zoos can be important in conservation education. A professionally planned exhibit such as the one at the Bronx Zoo teaches people about the animals, their habitats and problems facing them. For many people in towns, their first encounters with wild animals are often zoos. If these are properly designed with education and conservation as their primary aims, they can be important in showing people, many of them for the first time, the splendour of wildlife and value in preserving it.

The other role zoos can play for an endangered animal is to breed it. If a species is endangered in the wild, it can be bred in captivity and, at a later date, the offspring introduced back into the wild. This has been done successfully with a few species such as the golden lion tamarin, Arabian oryx and European bison. It is far from easy, however, and the problems are so great that it is rarely done successfully. For a start, the reason that the animals are endangered in the wild is likely to be because their habitat has been destroyed. In which case, there is nowhere left for the captive animals to be released. Another problem is that captive animals should not be released anywhere where there is a wild population, because there is a major risk that they might have picked up some infections in captivity which they then pass on to the wild population. This could devastate a whole population of an already endangered animal. An alternative is to put the

Generally, proboscis monkeys do not do well in captivity. They appear depressed, stop eating and soon die. This animal had been caught a week previously and was chained up outside a kampung shophouse.

captive-bred animals into areas where there are no wild ones. But that leads to the question of why aren't there wild ones there? Is it because they have been hunted out? In which case, there is no guarantee that the captive-bred animals will not meet the same fate. Or maybe there are no monkeys there because the habitat is not suitable in some way, in which case, released animals would not survive.

So breeding proboscis monkeys in captivity is fraught with problems, from obtaining the animals in the first place, to keeping them and re-introducing offspring to the wild. With extremely careful research and planning, it can, in certain very specific cases, be a highly useful conservation tool, but it is certainly not the widespread answer to conserving endangered animals that some people imagine. Only by ensuring that the animals' habitats are protected and hunting minimized can we ensure the survival of proboscis monkeys in the wild.

PROBOSCIS MONKEYS AND THE FUTURE

In 1964, the American zoologist J.A. Kern said that, due to the inaccessible nature of the proboscis monkey's habitat, and the fact that it was of little economic value, the monkey's future was not threatened. Since then, the picture has changed dramatically, and the proboscis monkey is now threatened throughout its range. There are two main problems: first, disturbance of the habitat, and secondly, hunting.

Generally, the places where proboscis monkeys occur are those which are most developed and inhabited by humans - flat coastal plains and riversides. Most people live along rivers, most large towns are near rivermouths, and most large-scale agricultural schemes are on the flat coastal plains. Such human pressure is the reason why proboscis monkeys are facing so many problems.

Mangroves, one of the proboscis monkeys main habitats, have been subject to intensive felling for their wood.

Loss of habitat

The proboscis monkey's main habitats are mangrove, peat swamp and riverine forests. Riverine forests are inevitably in thin strips, and are generally highly disturbed by the presence of towns, villages and by river transport. Few rivers in Borneo remain undisturbed, especially along their lower reaches where the monkeys are generally found.

In some areas, mangrove forests have been subject to logging so intensive that few trees are left. Some wood is used for firewood and charcoal, but generally such cutting is small scale and presents no great threat to the monkeys. Intensive cutting is done for poles for use in the local building industry, and wood and woodchips for export. There are only a few areas of mangrove left which have not been heavily cut in this way. Additional areas have been cleared, or are still under threat from clearance, to make prawn ponds.

This is in spite of the fact that mangroves are some of the most productive areas in the entire world, and are of enormous value to Man. So their destruction is highly damaging to Man's interests, as well as destroying the unique wildlife there.

The proboscis monkey's other main habitat is peat swamp forest. Peat swamps are accessible and full of valuable timber, so they have been heavily logged. If only a few trees are taken out, proboscis monkeys seem able to survive, such as in Tanjung Puting National Park, Kalimantan. In some areas, however, logging is heavy and followed by treatment of the forest to try to enhance the next timber crop. Treatment involves poisoning non-timber trees, many of which are good food trees for the animals. Thus, the food supply is devastated and the animals starve. This practice has largely been discontinued, but it has caused many animals to die in the past.

Between these various forms of habitat disturbance, there are few intact areas left within the entire range of the proboscis monkey. Each group of proboscis monkeys travels over large distances to find enough food, so populations need extensive areas of forest to survive. Few patches of forest, and even fewer reserves, are now large enough to include the entire area used by proboscis monkey groups during a year. Even if the monkeys only use food from outside a reserve for one month a year, they will die if the reserve becomes isolated.

◄ The conservation functions of intact mangroves and the commercial felling of mangrove stands cannot both be maintained without careful planning and knowledge of the dependence of wildlife on mangroves.

VALUABLE

Mangroves are forests which grow between the low and high tide levels, so they are inundated with salty water every time the tide is up. To keep themselves anchored firmly in the mud when the tide swishes around them, many of the trees have great networks of stilt roots above the mud. Apart from the seawater coming in every tide, mangroves are also fed by water coming in from inland rivers. These carry silt laden with nutrients, often from far in the interior of Borneo. As the rivers reach the mangrove, they slow down. The silt settles out and is trapped under the trees by the complex of roots. The nutrients are then taken up by the trees. When their leaves and fruits eventually drop, they are broken down by crabs and fungi to form a huge supply of food for fish, prawns and other sea-dwelling animals, both in the mangrove itself, and also way out to sea where the detritus is carried by the outgoing tide. In the Kuching Division of Sarawak alone, mangroves support fisheries worth more than M$50 million every year, and provide jobs for some three thousand fishermen. If those mangroves were to be cut down, all that valuable seafood and all those jobs would be lost.

Another thing that happens when mangroves are cut is that all the mud trapped by the tree roots suddenly washes out to sea and is deposited on sandy beaches all along the coast. This mud has been accumulating for hundreds of years. It is black, smelly and makes beaches extremely unpleasant. Again, in the Kuching Division of Sarawak alone, that would ruin tourist beaches that bring in a revenue of about M$20 million every year.

So from one small area alone, loss of mangroves would mean that an income of M$70 million would be lost every single year. Thus, mangroves are not only essential for their wildlife; they are highly valuable for humans.

MANGROVES

Hunting

In some of the few remaining areas of reasonably undisturbed forest, proboscis monkeys are hunted. Traditionally, hunting was not a problem because the people living in and around the proboscis monkey's habitats are predominantly Moslems, who do not eat monkeys. With the advent of speedboats and shotguns, however, people come from nearby towns and hunt for sport. The

Adult male proboscis monkey freshly killed by either a rival male or a predator.

proboscis monkeys' habit of sleeping conspicuously next to rivers every evening makes them especially vulnerable to hunters in boats, and large numbers can be killed in a short time. The number of proboscis monkeys in one hunted area in Sarawak dropped by 50% in five years.

The result of habitat loss and hunting means that number of proboscis monkeys is becoming alarmingly low in some areas. In Sarawak, there are probably one thousand animals or less in the whole State. These are divided into several small, isolated populations, none of which is totally secure. In Brunei Darussalam, the proboscis monkeys live in the mangroves of Brunei Bay, where they cross between Brunei Darussalam and Sarawak. They are hunted on the Sarawak side of the border, which reduces their population. There are probably 200 or less remaining. In Sabah, numbers are slightly higher because of the large populations in the Kinabatangan River and Dent Peninsula areas, so there could be 2000 to 3000 animals left in the State. This number is likely to drop rapidly unless steps are taken, however, because much of the forest is being logged for the second time, and large areas cleared for oil palm plantations. In Kalimantan, nobody knows how many proboscis monkeys are present because no surveys have been done. Numbers are undoubtedly higher than in northern Borneo because of the large areas of swamp forest and stricter control of firearms. They are certainly declining, however, due to loss of habitat in all areas.

Many of the proboscis monkey populations now are extremely small and far from others. No monkeys can pass between them. Small, isolated populations of any animal suffer from inbreeding, and are also very vulnerable to occasional disasters such as hurricanes, other major storms or disease. So many of the small populations will slowly die away. This is already happening in western Sabah and parts of Sarawak.

Conservation measures

Proboscis monkeys are protected by law in all regions of Borneo. This means that it is illegal to hunt or keep them or their parts, and penalties for doing so are high. Under international trade regulations (CITES, the Convention on International Trade in Endangered Species of Wild Fauna and Flora) the proboscis monkey is listed on Appendix 1. This means that it is illegal to move it between countries for commercial purposes. Export and import permits are needed, and should only be given if the animal is being moved for scientific or conservation reasons. Thus, on paper at least, the animal is well protected against being hunted, kept as a pet or traded. It is often difficult to enforce the law in areas riddled with waterways, however, especially when the wildlife authorities are extremely short of staff in all regions.

All regions of Borneo have systems of totally protected areas (TPAs) for protecting their wildlife and forests. All of Borneo's large reserves are in the interior, however, not in areas where proboscis monkeys occur. Coastal reserves are almost all too small to protect proboscis monkeys effectively; the monkeys

Another main habitat type, peat swamp forest, is also subject to legal and illegal logging. Here, logs are collected prior to removal by railway line.

are so wide-ranging, they need large reserves. There is only one TPA large enough to protect the monkeys effectively, Tanjung Puting National Park in Kalimantan.

Plans are underway to extend Samunsam Wildlife Sanctuary and Bako National Park in Sarawak, and to create a national park in the Lower Kinabatangan area in Sabah. If all three of these go ahead, there will be TPAs large enough to protect viable populations of the monkeys in Sabah, Sarawak and Kalimantan. That, combined with vigorous anti-hunting measures in all areas, will help to ensure that these unique and spectacular animals will survive for future generations to enjoy. If those plans are not implemented, with the current downward trends, the future for the proboscis monkey looks bleak indeed.

SEEING PROBOSCIS MONKEYS IN BORNEO

S ome of the places where proboscis monkeys occur are not open to the public, and others are remote and inaccessible. So we are not giving a total list of where proboscis monkeys occur, but a sample of the most accessible places where you can easily go and the chances of seeing the monkeys are high.

In all of these places, bear in mind that proboscis monkeys are most active in the early morning just after dawn, and in the two hours or so before dusk. These are also the times when the animals are alongside rivers, so plan your trips so that you are in the proboscis monkey areas at those times. With careful planning, and a little bit of luck, you will have a wildlife experience that you will always remember.

▲ The best time to view proboscis monkeys is just after dawn and just before dusk when they are gathered alongside the river.

▶▶ A dusk cruise along Borneo's rivers is an unforgettable experience.

Sarawak
Bako National Park

This small park is close to Kuching, and proboscis monkeys can be seen by walking extremely quietly on foot. There should be no more than two or three people together and absolutely no talking because this scares the animals away. Take the trails from the headquarters at Telok Assam to either Telok Paku or Telok Delima soon after dawn or before dusk and you stand a good chance of seeing the animals. Other animals which you should easily see are long-tailed macaques, silvered langurs and bearded pigs. The Park is also famous for its different coastal forest types, large numbers of pitcher plants, spectacular rock formations, sunsets and lovely beaches. To get to the park, take a bus from outside Electra House, Kuching to Kampung (= village) Bako. From there to the park is by boat, and there are always charter boats waiting at the Kampung Bako jetty, for a fixed price (M$25 per boat each way at the time of going to print). You can either take a day trip or, to stand the best chance of seeing the animals round dawn and dusk, stay overnight. A variety of comfortable hostels and rest houses is available at cheap prices. Booking is through the National Parks booking office, Sarawak Tourist Association building, Pangkalan Batu, Kuching. Telephone no.: (082) 248088.

▲ In some areas where there is hunting, proboscis monkeys are shy and flee rapidly; patience is needed to obtain a good view.

◄ With a little planning, it is fairly easy to see proboscis monkeys in all four regions of Borneo.

Pulau Salak Mangroves

This is also close to Kuching, but slightly less easily accessible. Take a bus from Kuching to Kampung Santubong or Semariang. Negotiate to charter a local boat from residents there. The best place to see the animals is the edge of the mangroves near Pulau (= island) Salak. Hunting has caused the animals to be shy, but your chances of seeing them are good if you go between 5 p.m. and dusk and cruise along the rivers and creeks in the Salak area. In any case, you will have interesting views of mangroves with their other unusual wildlife such as kingfishers, fiddler crabs and mudskippers. Sunrises, sunsets and rainbows here are often spectacular.

▲ Time spent watching an adult male proboscis monkey is an experience of a lifetime.

◄ All-male groups can look deceptively like harems, with one large male and several smaller animals.

Brunei Darussalam

There are no organised ways of seeing proboscis monkeys here, but it is easy to negotiate with the boatmen in the centre of Bandar Sri Begawan. They normally go to Kampung Air, but you can negotiate a reasonable price for them to take you out into the mangroves where you stand a good chance of seeing proboscis monkeys. The boatmen themselves probably know the best spots to go, but if they do not, try the twin small islands of Pepatan and Batu-Baru. Other mangrove wildlife and scenery and spectacular sunsets again are attractions.

Sabah
Lower Kinabatangan
The lower Kinabatangan region of Sabah is the best place in northern Borneo for spectacular views of large numbers of proboscis monkeys. The area has much other dramatic wildlife, including elephants, orang-utans, gibbons, red langurs, silvered langurs, long-tailed and pig-tailed macaques, bearded pigs, deer, hornbills, eagles, kingfishers - a wealth of some of Borneo's most impressive animals. Some such as proboscis monkeys and hornbills you can see from the river; to see others, you have to go into the forest on foot.

There are several ways of getting there, the easiest of which are:

1. Take a boat from behind the fish market in Sandakan to Suan Lamba. There are daily scheduled morning services, or else you can charter your own boat. At Suan Lamba, local minibuses and taxis take you by road to Kampung Sukau. At Sukau, charter a local boat to take you along the rivers.

2. Take a minibus from Sandakan, near the Community Centre, to Sukau. At Sukau, charter a local boat to take you along the rivers.

3. Charter a boat from Sandakan, across Sandakan Bay and down to Kampung Abai or Sukau. This is easy but much more expensive than the previous two options.

4. Take a bus from Sandakan to the Kinabatangan bridge. (The bus ultimately goes to Lahad Datu.) At the bridge is Kampung Batu Puteh where you charter a local boat to take you along the river.

5. By far the easiest option is to take a tour through a local travel agent. Several of the agents in Sandakan organise their own evening or overnight tours to see the Kinabatangan proboscis monkeys, and prices are usually very reasonable.

Lahad Datu mangroves
Proboscis monkeys occur in the mangroves near Lahad Datu. Charter a boat from Lahad Datu wharf to take you into the mangroves.

▶ Adult male proboscis monkeys are the largest arboreal monkeys in Asia.

Kalimantan

Kutai National Park, East Kalimantan

Go by road from Samarinda to Bontang. Go to the National Parks office there for permits and assistance with boat hire. Take a small boat to the mangroves around Teluk Kaba. This can be done as a day trip, or else you can stay overnight at the Park Headquarters at Teluk Kaba. Generally, there are good views of proboscis monkeys in the mangroves. Rehabilitant orang-utans at Teluk Kaba are an additional attraction.

Pulau Kembang and Pulau Kaget, Banjarmasin, South Kalimantan

From Banjarmasin, hire a local river boat ('klotok') to Pulau Kembang in the Barito River. The boat is very cheap (approx. US$3 per hour). Circle the island to see proboscis monkeys in the mangrove. If you disembark to see the Chinese temple, you run the risk of being mobbed by long-tailed macaques! No permits are required, and it can be done as an evening trip. For the more adventurous, take the boat further downriver to Pulau Kaget to see proboscis monkeys in the *Sonneratia* mangrove trees. This is a half- to one-day trip.

Tanjung Puting National Park

Fly from Banjarmasin or Pontianak to Pangkalanbun. Report to the police there for a letter of transit ('surat jalan'). Go by road to Kumai and obtain your permit from the Park Headquarters there. Charter a boat to Tanjung Puting. This is a regular tourist route and is easy to do. It is possible to stay at the new hotel at Tanjung Harapom. If you keep the small boat with you, you can spend several days exploring the Sikonyer Kanan and Sikonyer Kiri rivers. Apart from large numbers of readily visible proboscis monkeys, additional attractions are rehabilitant orang-utans and a wide range of peat swamp and lowland forest wildlife.

SUGGESTED FURTHER READING

If you have enjoyed this book and would like to find out more about Borneo's wildlife or about other primates, the following books might be of interest.

Cubitt, G. and Payne, J. (1990). *Wild Malaysia.* New Holland, London in Association with the World Wide Fund for Nature, Kuala Lumpur.

Davison, G.W.H. (1992). *Birds of Mount Kinabalu, Borneo.* Natural History Publications (Borneo) Sdn. Bhd., Kota Kinabalu & KOKTAS Berhad, Ranau.

Francis, C.M. (1984). *Pocket Guide to the Birds of Borneo.* The Sabah Society with World Wildlife Fund Malaysia, Kuala Lumpur.

Goodall, J. (1990). *Through a Window. Thirty Years with the Chimpanzees of Gombe.* Weidenfeld and Nicolson, London.

Kavanagh, M. (1983). *A Complete Guide to Monkeys, Apes and Other Primates.* Jonathan Cape, London.

MacKinnon, J.R. (1974). *In Search of the Red Ape.* Collins, London and New York.

Payne, J. and Andau, M. (1989). *Orang-Utan: Malaysia's Mascot.* Berita Publishing, Kuala Lumpur.

Payne, J., Francis, C.M. and Phillipps, K. (1985). *A Field Guide to the Mammals of Borneo.* The Sabah Society/WWF Malaysia, Kuala Lumpur.

Rubeli, K. (1986). *Tropical Rain Forest in South-East Asia - a Pictorial Journey.* Tropical Press, Kuala Lumpur.

Smythies, B.E. (1981). *The Birds of Borneo. Third Edition.* The Sabah Society with the Malayan Nature Society, Kuala Lumpur.

SCIENTIFIC NAMES OF ANIMALS AND PLANTS MENTIONED IN THE BOOK

ANIMALS

Arabian oryx	*Oryx leucoryx*
Baboons	*Papio* spp.
Banded langur	*Presbytis melalophos*
Bats	Chiroptera
Bearded pig	*Sus barbatus*
Chimpanzees	*Pan* spp.
Clouded leopard	*Neofelis nebulosa*
Colobus monkeys	*Colobus* and *Procolobus* spp.
Common chimpanzee	*Pan troglodytes*
Common langur	*Presbytis entellus*
Deer	Cervidae
Douc	*Pygathrix nemaeus*
Eagles and kites	Accipitridae
Elephant	*Elephas maximus*
Estuarine crocodile	*Crocodylus porosus*
European bison	*Bison bonasus*
Gibbons	*Hylobates* spp.
Golden lion tamarin	*Leontopithecus rosalia*
Golden monkeys	*Rhinopithecus* spp.
Gorilla	*Gorilla gorilla* or *Pan gorilla*
Hornbills	Bucerotidae
Kingfishers	Alcedinae
Leopard	*Panthera pardus*
Long-tailed macaque	*Macaca fascicularis*
Macaques	*Macaca* spp.
Nightjars	*Caprimulgus* and *Eurostopodus* spp.
Orang-utan	*Pongo pygmaeus*
Pig-tailed macaque	*Macaca nemestrina*
Pygmy chimpanzee	*Pan paniscus*
Proboscis monkey	*Nasalis larvatus*
Red langur	*Presbytis rubicunda*
Reticulated python	*Python reticulatus*
Silvered langur	*Presbytis cristata*
Simakobu	*Simias concolor*
Tiger	*Panthera tigris*

PLANTS

Figs	*Ficus* spp.
Legumes	Leguminosae
Nibong palms	*Oncosperma horridum*
Nipa palm	*Nypa fruticans*
Nutmegs	Myristicaceae
Palms	Palmae
Pitcher plants	*Nepenthes* spp.
Rambutans	*Nephelium* spp.

GLOSSARY OF TERMS USED

All-male group: Social group comprising males of different ages. Very occasionally, one or two females may temporarily join it. No breeding occurs in the group. It has a loose social structure, with animals joining and leaving it frequently.

Alluvial: Pertaining to the silt and soil deposited by rivers along their floodplains and deltas. This soil is rich in minerals and other nutrients, and plants growing there are generally highly productive.

Ape: Primate belonging to the super-family Hominoidea but excluding Man. Apes are characterised by having relatively very large brains and no tails. The lesser apes comprise the Asian gibbons. The greater apes are the common and pygmy chimpanzees and gorillas of Africa and orang-utan of Asia.

Bacteria: Microscopic one-celled plants. Many cause diseases, but some can be beneficial, e.g., those that ferment milk to yeast, and those in the stomachs of certain specialised animals (cows, goats, colobine monkeys) which help to digest the food.

Cercopithecine: Monkey belonging to the sub-family Cercopithecinae. They are found throughout Asia and Africa, and there are about fifty species. They are generalised monkeys with simple stomachs. They live in forests but, unlike colobines, some species also thrive in more open and man-made habitats.

Colobine: monkey belonging to the sub-family Colobinae. They are found throughout Africa and Asia, and are characterised by having complex stomachs containing bacteria which ferment the food. Apart from the common langurs of India, they are restricted almost entirely to forests.

Complex stomach: Stomach comprising a greatly expanded first section containing bacteria to ferment the animal's food, and a smaller, acidic lower chamber. The main advantages of the stomach are allowing the animal to digest cellulose (the main structural component of leaves) and to eat foods containing certain poisons without being harmed.

Genes: The basic unit of inheritance. The nucleus of every cell in an organism contains many thousands of genes which, between them, programme all aspects of the animal's form, functioning and certain aspects of its behaviour.

Grooming: Cleaning of the body surface by licking, nibbling, scratching or picking with the fingers. In primates, this has both hygienic and social functions.

Gunung (Malay): Mountain.

72

Habitat: The area and particular type of area in which an animal or plant lives in the wild.

Harem: A social group comprising one adult male, more than one adult female and their mutual offspring.

Home range: The area used by an animal or group of animals during a year.

Kampung (Malay): Village.

Macaque: Monkey belonging to the genus *Macaca*. There are 17 species of macaque, all but one of which only occur in Asia. They are generally brown, lively monkeys in large social groups. They live both in forests and in more open areas. They are intelligent opportunists with an eclectic diet, so many species thrive in agricultural and other man-made areas.

Mammal: Animal of the order Mammalia. Females of all mammal species produce milk to succour their young. Most mammals have four limbs and are covered with hair.

Mangrove forest: Coastal forest growing between the low and high tide levels. Since the trees establish in soft mud and are flooded with salt water every tide, they have developed many specialities such as stilt and aerial roots and the ability to secrete salt.

Nocturnal: Active at night.

Peat swamp forest: Forest growing on deep peaty soils. The substrate is permanently waterlogged and very low in nutrients since the only incoming water is from rainfall.

Prehensile tail: Tail which can cling onto branches and even support the animal's weight.

Primate: Mammals belonging to the order Primates, and comprising prosimians, monkeys, apes and Man. No one feature characterises primates, but they tend to have relatively large brains, agile hands and feet with nails, and good vision with stereoscopic sight.

Prosimian: Primate belonging to the sub-order Prosimii. Prosimians are generally regarded as the most primitive primates. They tend to be small and nocturnal, the only exceptions being the lemurs of Madagascar. Other prosimians include pottos and bushbabies in Africa, and lorises and tarsiers in Asia.

Pulau (Malay): Island.

Riverine forest: Forest growing adjacent to or near a river. In the tropics, it is characterised by being extremely dense and productive, and having large numbers of climbing plants.

Sexual selection: The process whereby certain feature(s) of an animal are preferentially chosen by a member of the opposite sex when choosing a mate. Thus, those features are passed on to future generations. With the passing of generations, if such features continue to be selected, they become exaggerated, such as the tail of a male peacock and nose of a male proboscis monkey.

Simple stomach: Stomach comprising only one chamber which is acidic and in which food is digested by enzymes produced by the animal itself. Such a stomach does not allow the animal to digest leaves or deactivate poisons, but it does allow it to eat a much wider range of other foods than an animal with a complex stomach.

Stereoscopic sight: Sight in which two eyes are facing forward so receive slightly different views of the same image. This gives an impression of depth, and allows the animal to judge distances accurately.

Stilt roots: Roots which come out of the trunk some distance above the ground and go into the soil. They give a tree a very firm foothold if the substrate is loose (e.g., mangrove mud). They might also trap silt in incoming water, which gives extra nutrients to the tree.

Tanjung (Malay): Peninsula.

Tannin: Astringent chemical produced by plants. It slows down digestion in animals eating the plant, so might protect the plant from being eaten too intensively. It occurs differentially in the plant, e.g., is generally in high levels in unripe fruits but much lower levels in ripe ones.

Totally protected area (TPA): Area in which any form of disturbance (e.g., timber extraction, land clearance in any form, building, hunting) is strictly prohibited.

PHOTO CREDITS

THE AUTHORS

Both authors were born in 1956, the Year of the Monkey: **Elizabeth Bennett** in London, UK, and Francis Gombek in Bau, Sarawak. Elizabeth then went on to Nottingham University to read zoology, and then to Cambridge where she obtained her PhD for research into the leaf monkeys of Peninsular Malaysia. It was this that brought her to Malaysia for the first time in 1978, living in the Peninsula for more than two years. After her PhD, she came to Sarawak in 1984, working for WWF Malaysia and Wildlife Conservation International of the New York Zoological Society (WCI) to conduct the first ever detailed study of the ecology of proboscis monkeys. This developed into a wider programme working with the Sarawak Forest Department to try to conserve the proboscis monkey and its coastal forests. She has been in Sarawak ever since. Elizabeth Bennett is a staff member of WCI and is working on a variety of wildlife conservation projects in the region, as well as training Malaysian wildlife scientists.

Francis Gombek had his schooling in Semunjan, Serian, Bau and Kuching in Sarawak before going to Universiti Pertanian Malaysia (UPM) in Peninsular Malaysia. After graduating with a BS in forestry, he did a short stint as a tutor at UPM in Sarawak before joining the Sarawak Forest Department in 1982. Ever since then, he has been in the National Parks and Wildlife Section. He has worked on many aspects of wildlife conservation, including parks interpretation, law enforcement, conservation education, wildlife surveys and inventories, and general parks and wildlife administration and development. He has been closely involved at all stages of the proboscis monkey work, both in the field and conservation planning. Francis Gombek is currently head of the Wildlife Research and Management Section in the Sarawak Forest Department.